Dedicated to you all reading this, thankyou all for your support.

To my partner, who puts up with me and my strange ways. For my 4 kids who make these stories come to life.
I love you all, thankyou

COPYRIGHT © 2021 BY NATASHA COOMBES

ALL RIGHTS RESERVED. NO PART OF THIS BOOK MAY BE REPRODUCED, COPIED, SCANNED OR USED IN ANY MANNER WITHOUT WRITTEN PERMISSION OF THE COPYRIGHT OWNER EXCEPT FOR THE USE OF QUOTATIONS IN A BOOK REVIEW. FOR MORE INFORMATION, ADDRESS: RILEECURTISAUTHOR@GMAIL.COM

FIRST EDITION

BOOK DESIGN BY NATASHA COOMBES

But how will we get there?
It's so far in the air

We could shoot up using a sneeze,
to send us up past the trees

Try jumping on a trampoline, So from far below we're as small as a bean

We need to get higher then that, higher then whats ever been mapped

Higher then a bird has ever flapped, even higher then Throwing your hat

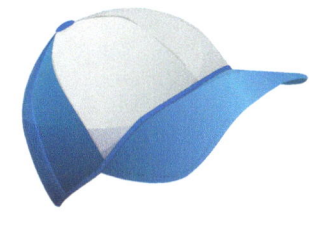

Let's go in this rocket,
The one in my pocket,
We'll make it bigger,
With this magic trigger

Why didn't you say this,
Now we can fly to the abyss,
To see all the planets that exist

They climb into the rocket,
The rocket from his pocket,
With the speed of a comet,
they neally vomit

Jupiter the one they can't miss,
The biggest planet in the abyss,

A gass giant, but the gasses are not from my bum,
Hydrogen and helium make this planet one,

The king of all moons, with 82,
they think I have more, but don't have a clue,
With around 30 rings,
Saturn is the king of kings!
Rings made from ice and rock,
They are like a mohawk

Uranus is an ice giant,
For this he is defiant,
The only one to spin, on the side,
With 13 rings, around they glide

Bright Blue gas giant is Neptune,
Orbiting me, there are 14 moons,

My winds are the strongest,
My 5 rings are hard to see,
But they are there trust me,

I'm the coldest planet out of them all you see,
I can reach about -221 degrees.

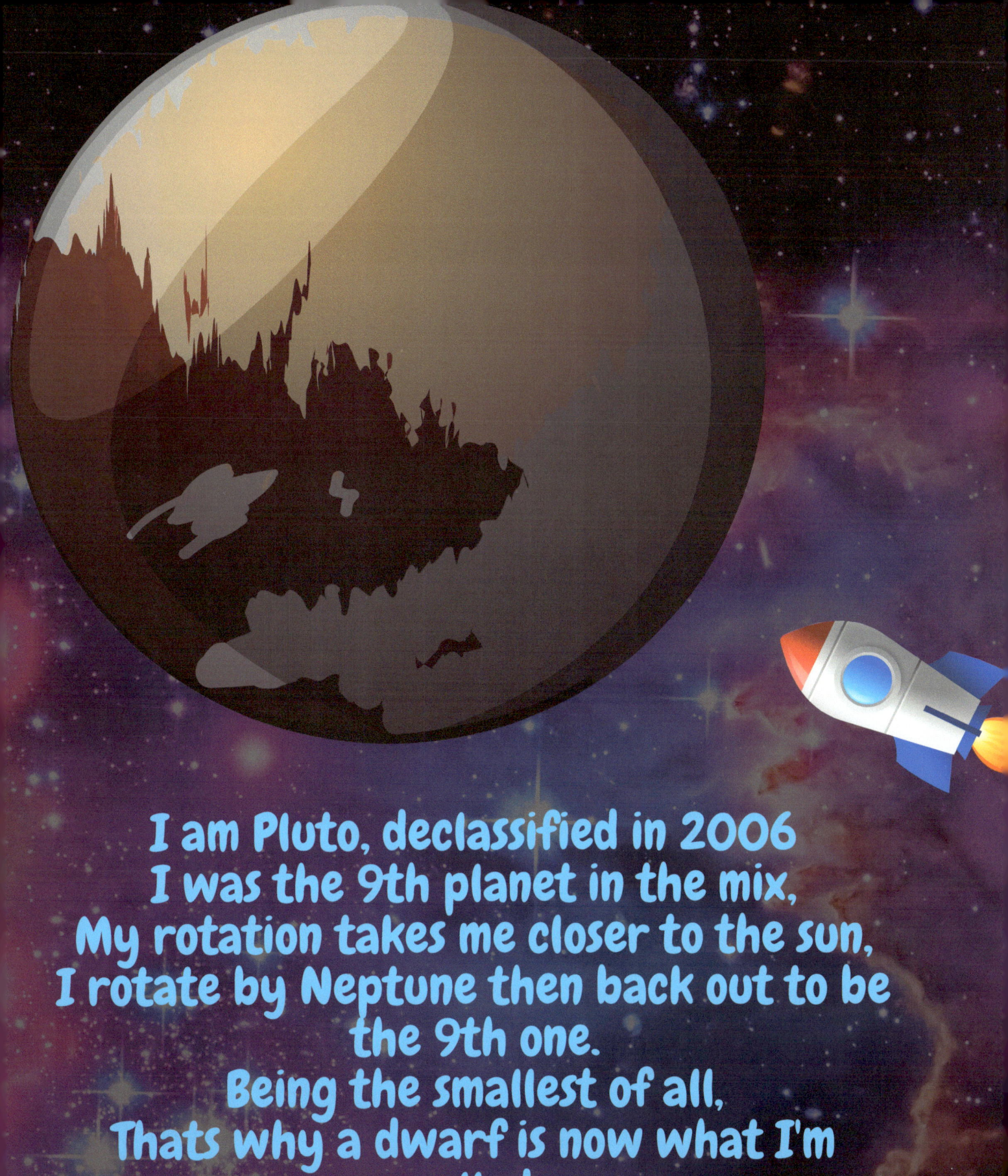

I am Pluto, declassified in 2006
I was the 9th planet in the mix,
My rotation takes me closer to the sun,
I rotate by Neptune then back out to be the 9th one.
Being the smallest of all,
Thats why a dwarf is now what I'm called.

SUN

MERCURY

VENUS

EARTH

Heading back towards Earth,
They realise that 2 have been missed,
The 2 closest to the sun,
Venus and Mercury are on the list.

2nd planet from the sun,
And the slowest rotating one,

Look up at the night sky,
Venus is the 3rd brightest to the naked eye,
Sometimes being that Bright,
Venus can be seen in broad daylight.

Mercury is the first planet from the sun, you see,
With no moons, and no rings,
The average temperature is 167 degrees,

A bit bigger than the Earths moon that you see,
To fill the Earth once, would take 18 of me

We have seen most the planets in the abyss,
There are more out there that were going to miss,
These are the main ones you need learn,
It's time for us to return.

Returning home, the rocket shrunk,
So he put his rocket back in the toy trunk

That was fun! They say, visiting all the planets in the milkyway

THANKYOU
For coming with us,
In our Pocket Rocket to the milkyway

I AM MERCURY

I MIGHT THE SMALLEST BUT I AM THE FASTEST

I HAVE NO MOONS

I GO AROUND THE SUN IN JUST 88DAYS CABSENCE II SPEEEDY

PURE VENUS
THE MORNING STAR

 2

 2

I AM THE BRIGHTEST PLANET, YOU CAN SEE ME IN THE SKY IF YOU KNOW WHERE TO LOOK

I'M VERY SLOW SO I DON'T GET DIZZY

I TRAVEL CLOCKWISE UNLIKE ALL THE OTHER PLANETS

I HAVE NO MOONS

LOTS OF VOLCANOES LIVE ON ME

KEEP EARTH CLEAN, IT'S NOT URANUS

MOON
GIVES US TIDES AND LIGHT AT NIGHT

EARTH
YOU LIVE ON ME I AM MOSTLY WATER

CLOUDS KEEP THE WARMTH IN, AND GIVES YOU PROTECTION FROM THE SUN

4 REACH FOR THE MARS
RED PLANET

YOU SEND ROVER THE ROBOT TO VISIT ME SOMETIMES

I LOOK HOT, BUT REALLY I AM COLD

I HAVE 2 MOONS

ANY WATER THAT I HAVE IS MOSTLY ICE

5 JUPITER
GAS GIANT

I HAVE RINGS BUT THEY ARE HARD TO SEE

MADE OF HYDROGEN AND HELIUM
"KISS MY GAS"

I AM THE BIGGEST

79 MOONS

7 CAN YOU SEE URANUS
ICE GIANT

THE BIGGEST MOON OF URANUS IS TITANIA

27 MOONS

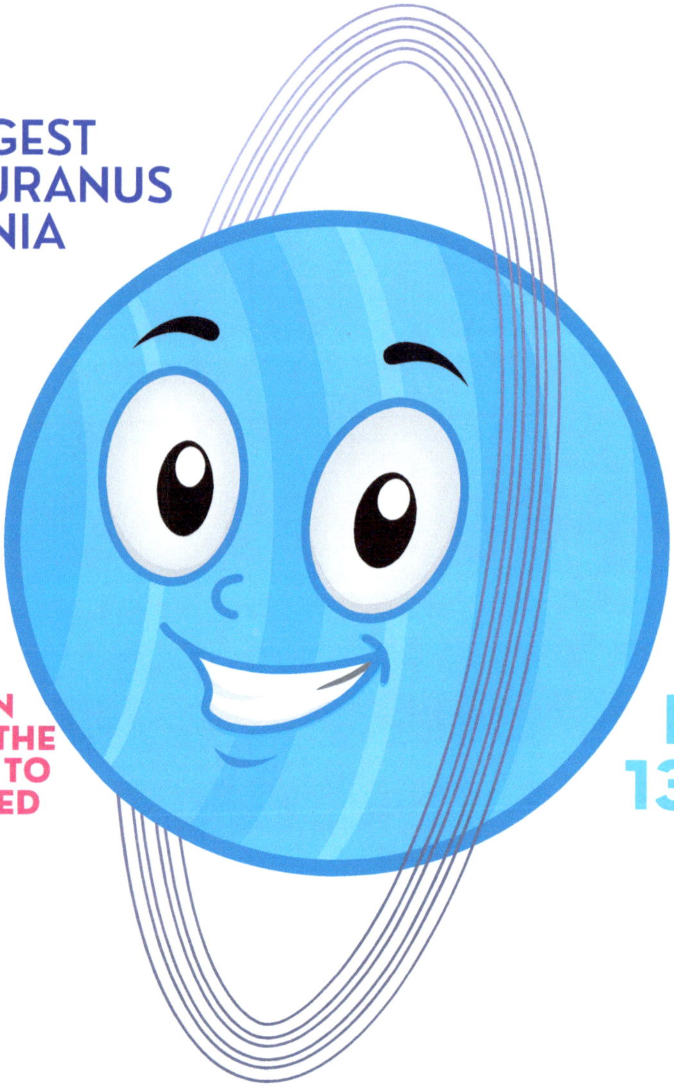

OTHER THEN EARTH I WAS THE FIRST PLANET TO BE DISCOVERED

I HAVE 13 RINGS

COLD PLANET -230°C/-330°F

I'M THE ONLY ONE THAT SPINS ON MY SIDE

8 NEPTUNE

14 moons

Coldest Planet
-221.45 degrees Celsius
(-366.6 °F)

Smallest of all the gas giants
hydrogen, methane, and helium

Largest moon (Triton) orbits backwards

165 Earth years, for neptune to make 1 orbit around the Sun

The strongest winds out of all the planets

9 PLUTO
DWARF PLANET

5 MOONS

01
1 EARTH DAY ON PLUTO IS 153 HOURS

1 EARTH YEAR IS 248 EARTH YEARS.

PLUTO WAS A PLANET FOR 75 YEARS BEING THE 9TH PLANET IN OUR SOLAR SYSTEM

PLUTO BECAME A DWARF PLANET IN  2006

My Vicious Earthworm Might Just Swallow Us Now

MERCURY
VENUS
EARTH
MARS
JUPITER
SATURN
URANUS
NEPTUNE

5 DWARF PLANETS
SMALL PLANETS

CERES

I'M BETWEEN THE ORBITS OF MARS AND JUPITER

PLUTO
I'M PAST NEPTUNE. I WAS THE LAST PLANET IN OUR SOLAR SYSTEM UNTIL I WAS RENAMED AS A DWARF PLANET

HAUMEA
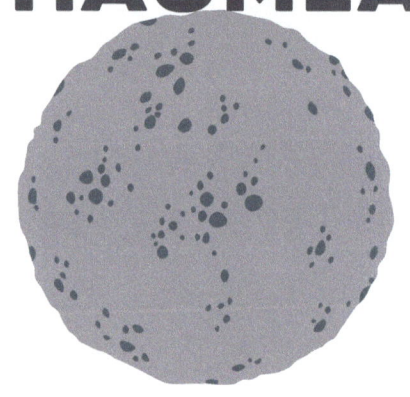

MAKEMAKE (MK)

MK2

ERIS
MY MOON IS CALLED DYSNOMIA

STARS

⭐ Most stars come in pairs "Binary Stars"

⭐ Small stars are RED
⭐ Medium (the sun) are YELLOW
⭐ Large stars are BLUE

The Sun is around 4.5 billion years old

The smaller a star, the longer they live

ASTRONAUTS
STAR SAILOR

The first Man to set foot on the moon was American Neil Armstrong

Astronauts have to learn Russian

Astronauts wear nappies

Animals were the first to be sent into space

Monkeys, fish, mice, dogs and bees have all been into space

SPACE ROCKETS

Can travel as fast as 15,000 mph in 8 seconds

Biggest rocket is over 300 feet tall

Rockets were first used as weapons

The sound of a rocket is sooo loud, it could actually destroy the rocket

The system uses water to absorb the sound as the rocket lauches

Almost as loud as a nuclear bomb explosion

PLANET JOKES

Q. What is an astronaut's favorite key on the keyboard?
A. The space bar!

Q. Why did Mickey Mouse go to space?
A. To find Pluto

Q. What planet is full of cows?
A. The MOOOOOn!

Q. How do the aliens get their baby to sleep?
A. They ROCKET!

Q. What do planets sing?
A. Nep-tunes

Q. What star wears sunglasses?
A. A movie star!

Which planet is the richest of them all?
Saturn, because it has many rings

www.ingramcontent.com/pod-product-compliance
Lightning Source LLC
Chambersburg PA
CBHW040410220526
45473CB00004B/1187